Saturn surrounded by three of its many moons

Saturn

Steve Potts

A⁺

Smart Apple Media

COPYRIGHT

Published by Smart Apple Media

1980 Lookout Drive, North Mankato, MN 56003

Designed by Rita Marshall

Printed in the United States of America

Pictures by Photo Researchers (Julian Baum/Science Photo Library, Chris Butler/Science Photo Library, Lynette Cook/Science Photo Library, David Ducros/Science Photo Library, Library of Congress/Science Source, Steve Munsinger, NASA/Mark Marten, Science Photo Library, U.S. Department of Energy/Science Photo Library), Tom Stack & Associates (JPL/TSADO, TSADO/NASA)

Library of Congress Cataloging-in-Publication Data

Potts, Steve. Saturn / by Steve Potts. p. cm. — (Our solar system)

Includes bibliographical references and index.

ISBN 1-58340-098-2

1. Saturn (Planet)—Juvenile literature. [1. Saturn (Planet)] I. Title.

QB671 .P68 2001   523.46—dc21   2001020126

First Edition 9 8 7 6 5 4 3 2 1

Saturn

A Giant Planet

Saturn was the ancient Roman god of agriculture. He symbolized growth and prosperity. It is fitting then, that the second largest planet in our solar system was named after him. Amazingly, Saturn is 800 times larger than Earth, yet it would be light enough to float if it were dropped into an immense ocean. ☀ Before the invention of the telescope, Saturn was the most distant planet that people could see. The Romans thought the distant yellow object was a "wandering star"

An illustration showing the planets' unique coloration

because every night they saw it in a different place. ☀ With

the invention of the telescope, astronomers could see Saturn

more clearly. They were amazed at what they found. The first

scientist to see Saturn with a telescope was **Saturn's day is 10.2 Earth hours long; its year is 29.46 Earth years long.**

the Italian astronomer Galileo Galilei in 1610.

He thought that Saturn was three separate

planets. ☀ In 1655, astronomer Christian

Huygens discovered that Saturn only looked like three planets

because it is circled by rings. Three other planets—Jupiter,

Uranus, Neptune—also have rings, but Saturn's rings are the

only ones that can be seen through a small telescope.  As

telescopes improved, astronomers were able to find out more

about these strange rings. In 1970, American astronomers dis-

Galileo Galilei, the first man to view Saturn closely

Saturn's rings are made of small pieces of rock and ice

covered that Saturn's rings are not solid. They are actually billions of pieces of rock and ice. Some of the pieces are as small as a snowflake, and some are 10 yards (9 m) in diameter.

Saturn's Characteristics

Saturn is a long way from the Sun—890 million miles (1.4 billion km)—so it is very cold. Temperatures in the clouds around Saturn reach only –279 °F (–173 °C). ☀ Huge storms batter Saturn's surface. These storms travel across the planet and produce big white spots near Saturn's equator. ☀ Much of Earth's atmosphere is nitrogen and oxygen. Saturn,

though, has an atmosphere that is 97 percent hydrogen, 3 percent helium, with trace amounts of methane and ammonia. It would be impossible for humans to breathe Saturn's air and

Two of Saturn's moons in orbit around the planet

survive. ☀ Because Saturn spins so fast on its **axis**, it has

been flattened at its poles. It looks almost like an egg that has

been laid on its side.

Moons and Rings

Saturn has six major rings that circle its equator.

When viewed closely from space, these large rings divide into

many smaller rings. The rings may have once been planets that

were captured by Saturn's **gravity** and broken into pieces.

Through a telescope, three of these rings show up as thin but

very bright circles. In photographs taken by spacecraft, it's pos-

sible to see right through the rings. ☀ In addition to its rings,

Saturn also has 18 moons—more moons than any other planet.

Some of these moons are very small. Pan, a moon that is close

Saturn's moons range in size from tiny to gigantic

to Saturn, is only 12.5 miles (20 km) wide. Five of the small

moons were discovered in 1979 and 1980 by powerful tele-

scopes and *Voyager 1*. The *Hubble Space Telescope* recently

detected four more objects that appear to **The dark gap dividing Saturn's rings into two parts is the Cassini Division.**

be moons. ❋ Saturn's Titan is the second

largest moon in our solar system. Titan is

even bigger than the planet Mercury. It is so

big that it was discovered by telescopes in 1655.

A drawing of a probe exploring the rings of Saturn

Exploring Saturn

In 1979, probes began to fly by Saturn. Probes are small

spacecraft that are sent into space by rockets or are carried on

a space shuttle. These probes carry complex computer and

photographic equipment that is used to take pictures to send back to Earth. Much of what we know about planets like Saturn comes from probes. ☀ *Pioneer 11* was the first probe to fly by Saturn. It took the first photographs of the planet. From 1979 to 1981, *Voyager 1* and *Voyager 2* flew by Saturn again. These probes provided information about Saturn's moons and rings. In 1997, the *Cassini* orbiter was launched. It will reach Saturn in 2004 and **orbit** the planet for four years.

Every 30 Earth years, Saturn has a massive storm called the Great White Spot.

The *Cassini* orbiter approaching the moon Titan

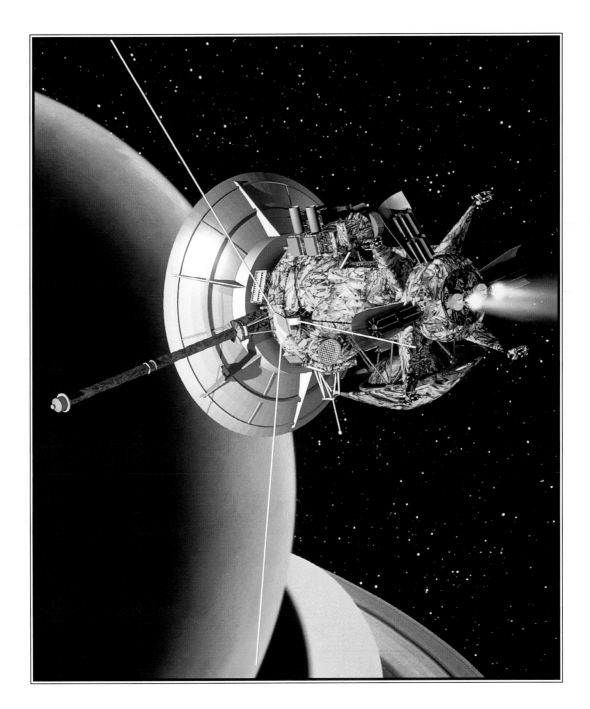

It will drop the *Huygens* probe onto the surface of Saturn's moon Titan for a closer look at the planet. ☼ Once scientists start receiving information from *Cassini* and *Huygens,* they will learn even more about Earth's distant neighbor. Just like early astronomers who peered into their telescopes to view this giant planet, modern scientists are using technology to uncover Saturn's secrets.

Saturn's outer ring stretches 300,000 miles (480,000 km) from the planet surface.

An illustration of Saturn as seen from Titan's surface

Saturn's rings may be the remains of an exploded planet

Index

Words to Know

atmosphere–the nearly invisible layer of gases that surrounds a planet

axis–a non-moving, imaginary line that an object rotates around

equator–the imaginary ring around the center of a planet that divides the northern half from the southern half

gravity–a force that attracts all objects in the universe; it's the force that makes things fall to the ground

orbit–travel in a repeating circular pattern around another object

Read More

Bond, Peter. *DK Guide to Space*. New York: DK Publishing, 1999.

Couper, Heather, and Nigel Henbest. *DK Space Encyclopedia*. New York: DK Publishing, 1999.

Furniss, Tim. *Atlas of Space Exploration*. Milwaukee, Wisc.: Gareth Stevens Publishing, 2000.

Internet Sites

Astronomy.com
http://www.astronomy.com/home.asp

NASA: Just for Kids
http://www.nasa.gov/kids.html

The Nine Planets
http://seds.lpl.arizona.edu/nineplanets/nineplanets

Windows to the Universe
http://windows.engin.umich.edu